IDOLS

STEPHEN ROMER

IDOLS

for Isa,

with thanks for
buying this book!

Stephen Romer.

London 9 . VI . 88 .

Oxford   New York
OXFORD UNIVERSITY PRESS
1986

Oxford University Press, Walton Street, Oxford OX2 6DP

Oxford New York Toronto
Delhi Bombay Calcutta Madras Karachi
Petaling Jaya Singapore Hong Kong Tokyo
Nairobi Dar es Salaam Cape Town
Melbourne Auckland

and associated companies in
Beirut Berlin Ibadan Nicosia

Oxford is a trade mark of Oxford University Press

British Library Cataloguing in Publication Data
Romer, Stephen
Idols.
I. Title
821'.914 PR6068.04/
ISBN 0-19-281984-4

Typeset by Joshua Associates Limited, Oxford
Printed in Great Britain by
J. W. Arrowsmith Ltd., Bristol

for Biddy, and for my Parents

ACKNOWLEDGEMENTS

Some of these poems first appeared in the following publications: *Firebird 4* (Penguin), *Harvard Advocate*, *New Statesman*, *PN Review*, *Stand*, *TLS*, *Twofold*, *Word and Image*.

'Higher Things' was awarded the Prudence Farmer–*New Statesman* award, 1985.

'Wellingtonia' was runner-up in the *TLS*-Cheltenham Festival Poetry Prize, 1985.

CONTENTS

PART I

'She's only
a circumstance of self that nothing reconciles.'

W. Logan

The French Translation

A copy of *Portrait* by her unmade bed . . .
Embracing in their common hatred

what am I, against the gut alliance
of Catholic Ireland and Catholic France?

But Dedalus, I know you through and through!
We even share a name. Reading you

I sprouted wings and fled. We are both
at an angle to England, travelling south.

Will you, this once, speak for two of us,
direct her simple wilful heart, release

those channels to remorse, possess her mind,
as I come flying humbly on behind?

I doubt you would enter in so far.
What were your ardent ways but a posture

for being in despair. You had the knack
of detaching what you needed from the ache

of merely needing . . . Her brief, stifled yawn
has frazzled my patchwork wings to the bone.

. . . I glimpsed you as I fell, you venerable
heartless survivor, flying out of trouble.

Remedies

Learn the use of unhappiness, recommended
apostolic Proust on his death bed,

and no emotion will go to waste.
It's a tempting feast,

but I shall not be drawn a second time
to her forcing-house of a room

where I showed my hand too soon,
those premature, acid green

wrinkled chestnut fingers, unexplained
along her street. I shall remain

at liberty to interpret
the ambiguous spirit

of her calling from the blue like this,
which alone is happiness,

a prolonged underglass imagining
for the asthmatic in his scentless spring.

Grammar

I stick to facts and teach the rules of grammar.
Pasts both perfect and simple, the clamour

of auxiliaries, common defective verbs.
Beginners are secure. Nothing disturbs

their present world with its concrete nouns
in place, as I lead them up and down

a serviceable house. Later I teach
the troubling losses of reported speech

in which my present now becomes his then
and what I say he said, without the passion,

as the active verb I saw can soon
decay into the passive she was seen . . .

The early blue is my example
of the limpid future simple,

as when we say it will be fine,
before the rapid clouding through the pane,

the change of tense, the spoiling of a life.
Your heart would have responded, if . . .

From the Corner Seat

There is comfort in tables, and equilibrium
in a corner seat. Aromatic steam

. has taken the place of my head,
it might be fumes across a mask of the dead

I have sat so unnaturally still
this last hour, probably with a smile

that I think might even be wisdom
if wise it is we become

from sitting unnaturally still
and letting the intolerable

be changed into an unimportant detail.
I can drink what she drank, inhale

her brand of cigarette,
and it's easy to interpret

the skull contracting against the skin
—any second she might walk in . . .

Our café is the same. She's as good as here.
There's life in a table and an empty chair.

Evil Counsel

Hunched in a coat in a corner
I can see the street in a mirror

and the street itself as my enemy
comes smilingly towards me

from two directions. I could gulp my beer
and vanish, if it were an answer,

if hell were merely other people
who are, with practice, avoidable.

But hell is what we can't control,
the evil counsel

he whispered in her ear
when I was nowhere

not even in the mirror.
Now she's stepping blithely through a door

and what I need, though I sit in terror,
is the neutral man who can measure

just how close this is to love
by watching its object move

away from me in two directions.

Brasserie Lipp

The back of my head in an angled mirror.
Unflinching self-regard starts here,

with a disappointing skull. It falls away
from the crown, quite unceremoniously

finished off, behind a good façade.
There could be worse, something overbred,

the line of least resistance ... To a girl
who's just sat down, I bare my Aztec profile.

Her skull is shapely but her lips are tight,
her eyes disdainful, but on the lookout;

imagine my surprise when she caressed
so tenderly the waiter's hairy wrist

bent like a croupier's to rake her money in.
It's a problem of interpretation.

Her gesture darts into four bright angles,
—infinite self-advertisement? Or else,

but there's a rhetoric to this as well,
'the shocking loneliness of the beautiful'.

The Flaying of Marsyas

The fool who played his pipes too well.
Ovid recounts it, in surgical detail,

how that nimble-fingered ladies' man
was separated from his skin

for thinking he could play two pipes at once.
Now put a twist in the tale, an alliance

between the courted pair. Net him for society.
The women I love met up today

for tea and Italian art. A chaste conversation.
Profane Love reclothed for a Visitation . . .

Titian strung Marsyas upside down;
he hangs apart from the two strong women,

cut out, unremarked, left to play the martyr.
An ardent, undignified, whining satyr.

Theory

Every point of view is described by a circle.
I'm alone in the house, my Prospero's cell,

leaning to the peevish flame of an illusion
that's circular, vicious, and admits no-one.

Which geometry is real?
There's one that says we live in parallel

and never touch. That we share, at best,
worsening November weather, the common frost

that struck last night, and international
catastrophe on the news. It is a trivial,

piercing hypothesis . . . Here, I try to keep warm,
shuttling from the coal shed to my room,

and dream like Caliban of getting free
by breaking magic wands across my knee.

Higher Things

I wish I could, like Søren Kierkegaard,
be absolute and let her face recede

until it is an island in the water
he called memory. Nothing impure

could touch his lasting image of Régine.
Only in memory is love immune

from longing to be with her all the time.
He kept a candle burning in each room,

unfinished manuscript on every desk.
I shall need all his courage for the task

of settling firmly to the sublime;
there is only her face to start from.

Alternatives

If not the white night of Propertian
triumph, at least let me mourn,

singly through the small hours, for lack of it.
The seamstress ravels up her strands, but not

with me in mind. There's a sick perspective
of light and dark, measured out and wasted in a sieve.

Defiance then. Grant me a storm,
a night of *Deum Meum*, *Deum Vestrum*,

lightnings and a long renunciation.
Content with that, I'll fall asleep at dawn.

Theology of the Flesh

Niceties of the Trivium,
sic et non. And may his God forgive him

for doing what he did to that girl.
He hurried her into the veil

and forgot her. His Sister in Christ.
She thought he put all to the test

'and found her bitter', for she had no word
from husband, master, especial lord,

Brother in Christ, what could she call
a man reformed and unimpeachable?

The worst of it, in time she came to sing
in praise of death, his happy ending . . .

A slippery logician, bred
in the schools of Paris. The type dies hard,

for she has her's, an elder schoolman
ripe for the inferno . . . Soon

I shall remind her, if need be on my knees,
how Abelard put away his sage Heloise.

Carmen

She is in black, but here despair is red,
her lips saying No, and the bloodied

bull-cape coming back . . . I saw the film alone,
free to indulge grotesque comparison

and come out shaking. She sprinkled a circle
about their bed, and tossed him a smile

that landed round his neck like a lasso.
She pulled it tight with a word like Go,

and a sulk that yanked him back to her
as down she went all tenderness and laughter.

Timing faultless, she broke a plate and danced
which set up heavings in the audience

that sickened quickly as our hearts went out
to little José when she looked him straight

in his cloudy eyes, like handing him a rope.
He was upright but fell to unmanly worship:

it elicits no reply. Cryptic as Christ,
she drew a single figure in the dust.

After Corbière

I

We're in Paris, child; learn to deny
your dismal, po-faced colony,

its marshlight and belltower.
Youth does end, its end is here.

It's time you scrapped your overwritten
early loves and learned to poke fun

at the real thing. Your lyric cup
is full, but throw it out and keep

the dregs, the bitter ones, like these . . .
A harmless trick, and you'll be pleased

when the literal-minded upright man
thinks of you and smells corruption

—or does not think of you at all.

II

I loved—but that won't sell:
if you want to get deep in your girl

you have to dig deep in your pocket . . .
My lover said 'I shan't forget',

and now in this scent of lilac
she ghosts me till I'm sick

with it, and does she weep as well
perhaps? It matters little,

the loss is mine; I'll lie awake
to sing it, uncandled and insomniac.

I'll pen my sad matutinal verse . . .

but here and now I shall disguise
my reddened eyes

and have her whorish, orgiastic, fierce.

III

You laugh. We've sunk to bitterness then.
I'll learn some mephistopholean

lip, foam it with absinthe, and call it wit.
I'll say it comes straight from the heart.

I'll bowdlerize love—one boredom less—and leave
it on the shelf. Let posterity receive

the gift of myself . . . Even now my scabby lung
breathes the stench of glory, the victor on his dung.

Enough of this, I think. Now move on,
but leave your wallet and your weapon,

my ultimate mistress and final friend.
A curious pistol to make an end

of life . . . Or otherwise I could stay
at a sordid table and drink it away . . .

IV

You're a poet. So what? We need more
than that: Parnassus up an escalator,

a sexual-cum-religious freak,
a cripple or 'lovable eccentric' . . .

The Difficult Poet bathes in his difficulty,
elbows on the zinc in the 'Poison Tree';

for the Innocent a rose is a rose trala
and his spade a spade tralee:

'why can't people say what they mean?'
But his metre's a ball and chain . . .

The rose is a rose in antiquated Nature;
the art's a bonus, practised by my pedicure.

(I–IV from 'Paris')

V

To a Milky Youth

Iambic youth, you put your oar in everywhere
and heave away, mostly shipping water;

you've left out nothing but what you need,
your verse is green, unseasoned salad.

Try the ladies for remedy and target:
they'll tune your nerves to plangent gut

and puncture your floating verse balloon.
You'll crumple into truth when left alone,

a master of the nervous instrument.
Lower your Muse and be content.

Now you're a sucking pig, out at grass,
pink and soft and ignorant. The curse

comes later, with the salt taste of sex,
and when you're withered up, in specs,

you'll scan that innocent sepia book
of juvenilia with a famished look.

Of Comfort in Books

Books. Can they help? Is it consoling to know
his love had a pattern mine may follow?

'Triumphant progress, Brumaire to Floréal.'
In Paris, too. But he had circumstantial

perks, a Revolution fairly made his name.
To his natural wit we can add the smell of fame.

The months have lost those lovely names and we're back
in the Ancien Régime, she and I, stuck

in somebody's *Amours* at chanson seven.
From here to chanson twelve is hell to heaven

but our game old poet seemed to work his miracle.
Or was he lying, to finish his cycle?

News of Her

Because the world is wider than my thought
and reasonable voices tell me I'll forget

tonight I paid attention to the news.
Another tanker's stuck off the Azores

where I'm told it 'foundered' and 'caught fire'.
How much of my sympathy can this require

if the captain has been rescued with his crew
and the blaze will be dead by tomorrow?

But if one poor soul had been left on board
he would have made me pray (my thought is broad)

for the many who founder and catch fire,
for the few still burning, six months later.

Resolve

With the problem of the jeune fille
the cosmos and my life to reconcile,

I sat down gravely in a neutral bar
to chew the bread of things as they are

and drink the bitter blackest coffee
ever served by Dame Philosophy,

the grey mistress of severities,
keeper of the pincers, ought and is,

that hold us wriggling in our condition.
Be practical, she said. My consolation

is in your ecstasy when you abandon
hope, and there's nothing to be done.

As for the girl, forget her
and start afresh. A minute later

I tried this out: starting now, I am made
anew. Starting now, I shall conclude

this matter, pay my bill and enter life.
Starting now. It cannot be put off.

—One more coffee then. I shall prevail,
resolved at last. There's just a detail:

am I changed or not, if I tell this tale?

Steam

I siesta beneath a rampart wall
where Jacob's ladder is a horizontal

skytrack, fading into rungs of smoke,
a blue sublimation of our walk

on a disused branch-line sleepered in the plough,
an orange gash in the violet shadow,

settled between embankments. With rust and shine
on the tracks, we tried to imagine

how those railway builders could endure
the straightness of their lives and not desire

to strike out over fields at an angle
to the line, in revolt at the single-

minded purpose of the thing ... Turning livid,
the sky was criss-crossed with trails overhead,

a network of alternatives, the lines
in a used palm, our minor temptations

corroding in electric blue, fabulous
ladders runged in cloud, where others pass

with ease up and down, amorous and free
to alter, though not substantially

like us, who toe the line between steep
banks, slightly hobbled, more and more in step.

The Allotment Path

I like this path of slime, or baked mud and flint,
how it serves each man in his allotment,

going equably its way, a dogged strip
between two lots of vegetable and scrap,

the proud domains, behind the chicken wire
twined with barbs and parodic creeper.

A man in his shanty, a cockerel
on his crowing post. He gives me hell

in querulous falsetto, and drops down
to the dung-hill harem where his claws sink in . . .

Retrenchments. But the buildings opposite
seem a patchy row of housefronts, a stage-set

in the desert, trimmed with orange branches,
the willow's spiky flame, or the pruned inches

of its reddened life. The water
in the marsh is hosting a fire

that feeds untended like jealousy or wrath,
choler on black bile. I keep to my path

until it narrows discreetly and stops
at an iron fence where the ground slips

downward into slime. I'm afraid it led
nowhere in particular . . . Just to one side

in the grass, there's an upright cabin,
a tall outhouse for single meditation,

and after the undeluded death of friendship,
I like this broken *chiottes* on the ash heap.

Entre Chien et Loup

(for Gilles Ortlieb)

Evening is halogen and cobalt, hunger
and nerves, an objectless desire,

jealousy, enthralment, freedom;
small disturbances before the train home.

And the word 'attrition' is the smell
of oxidized sulphur in the tunnel

of disguised pornography.
We are haggard with frustration, and the sky

has gone more wolf than dog in the interim . . .
I glide into a bar and out of harm.

In the 'Rendezvous des Belges' there's a man
my veering mind does well to rest on;

a man of substance, with his leg braced back
and his elbows on the zinc, a man of stomach

and slow time, savouring a lager
from the flatlands. They are north of here,

flinty plains and market towns, where the sun
at evening is an inflammation

that lasts for hours. The place was home
for Emma Bovary, a flawless boredom

on which her little sulks were lost,
and all her niceties turned to dust.

She was useful as a scarecrow,
an elegant figure in a field of plough

craving *Passion*! *Félicité*! *Délire*!,
a steamy-shrill arrival at the Gare du Nord

where my stick-in-the-mud will take his train.
I watch him check and pocket his change . . .

Exulting in that huge equilibrium,
I get up when he leaves and follow him.

What You Will

She was radiant and alone on Garda
— mais ça ne vous regarde pas —

which is true enough, it never did or does;
it is godlike, nonetheless,

to watch her from a lakeside villa
up to her neck in buoyant water,

—to be that water, then her towel,
then the waiter in her small hotel

and cling to her all night as sheets . . .
The game is useless which has no limits

and the mind can free itself from anywhere.
Its laws are written on running water

and the water feeds a lake I've never seen.

South

Becalmed in a thirsty garden
with a fig tree, I might yet learn

the sybaritic virtue of forgetfulness,
and how sweet Muscat in a glass

embodies wisdom. Or I could emulate
the drier palate

of a Roman prince of pastoral,
have him share my pauper's meal

of bread, tomatoes, olive oil,
fix my bilious fear of alcohol

and ceaseless blue, wrinkle my skin
to Mediterranean

lizard, deftly untangle
this Gordian matter of a girl

and lay it open in the sunlight
with his candour . . . 'When you're written out

it shows. Call a halt, decant the best
of your intensity and love its taste.

Your vision of the girl will never fade;
it is neither comfort nor vicissitude.'

PART II

'I have amended my life, have I not? he asked himself.'

James Joyce

Sea Changes

The novices drink and smoke, or stand
on deck, blinded by the rhetoric
of a riding moon with clouds. Our wake
is white, a crumpled parachute
spreading out behind.

Grown used to this journey through the night,
wrapped in a coat, curled on a seat,
I ask only for a heart as constant
as the throbbing of this ship, and strong
for each new sickening of the sea.

How Things Continue

A sheet hung out of a window and shaken
on a morning of sunlight and warm bread
will gather such lights into its folds
that it drains the street and all you see
is this radiant flapping thing, until
it is as suddenly withdrawn, as if there
had been nothing but a black window space.
So the intricate green crystal suspended
at the end of an avenue in autumn
dissolves on approach, and each small light
goes out on its blade.
                              It changes nothing;
the milk heats in its pan, and sunlight
breaks on the faces of my friends. Only
at times, but more and more, these absences
and a voice: 'that things continue as they were
is thus the more dangerous and terrible'.

Aspects of this Sickness

All week the skies have moved upon despair
requiring through folds of various grey
my alteration. They will not allow
a staying place, but move continually
to draw me from my strangling circle,
pierce me with the cry of a bird
and loose upon my head the rains
of immediacy. Those effortless
fallings! No willing brings them down
or spirits away despair.
                              But one
despair is spirited; it will move
on workable acres, harrow and turn,
in slowness lift, and over a field
of yellow mustard set the grey skies flying.

Since I Woke

Brought to life from the sleep of children
they sprawled on the mountain, bewildered
but changed, speechless and
answering . . . I stay the same
lying on grass, noticing a cloud
that stopped behind the beech and clothed it
with midsummer mass, shone out and passed on
to leave a tree unchanged.
           Since I woke
to find the law is green, the sun long up
and swinging over acres, I am still
what most I was, a man scaled with sleep,
an offence to all the wind and light,
of every gust the backward cry
I am not yet ready! . . . It is too late!

Thinking on these Things

Lines of a winter tree, maintained in light;
a lovely responding girl; these are the things
that wound. They recall a place, not here,
where hour after hour the blood seeped up
from the soil of little fields, until it streaked
the chaffinch on its branch; a place where hands
were quiet and did not clench, where a touch
was enough, with no word said.

       It is faded,
and each lightless day I hear the crying shades:
'only think of us, with our fluent speech
and thin blood, who have lingered all our lives
on the foothills of assent; we in whom
the ineradicable sown-in-us
might on the instant there and then have grown!'

There We Have Been

These are the unbreathing days, blocked
from end to end by neutral sky, when
desire is to force it open and fling
a window wide on to trees on fire . . .
We have stood behind it, in a room
where all was changing and accomplished,
your body annihilated against
the fiery wall by the light it sponsored
leaving the disk of your face, an icon
that lived . . . The folds in a cloth
or apples in a bowl would conjure up
the ghost of some shy master
until it vanished, and these last days
we are dispossessed, and where we lived
might not exist, dwellings in a furnace.

Something More

Unceasing voice, old familiar, what have
you to say to me now? I have listened
to your murmur and obeyed. The time
is past for promises that always seem
to be for all, then nothing. 'Commune with
your heart, in your chamber, and be still' . . .
At times the only moving thing is sky,
a square of changing colour, blue or grey,
most often white. Clouds pass through,
brief tenants in the window opposite.
These days too similar were mirroring
a solitude too absolute, when over again,
voice of exhortation, you begin,
as if there were something more to say.

Wheels and Laws

Within the hollow middle of a day
the god of small surprises lies in wait
to open up on any wall an eye,
a little geometer's work of light
revolving round and round. Only the stray
and spacious mind will ever notice it,
falling silent to hear the cry
of someone lost between two thoughts,
and found again before he goes his way.

In deserts of the law, down a gauntlet
of narrow saints, unable to deny
nor yet confirm, my eyes were caught
by the flowering of the blue, where the sly
god sits in a spinning wheel of light.

This Poem for Burning

'I founded my house on meaning, the extreme
that knelt in yellow to the west, or spoke
from a simple twist of light, an arrow
pointing upward on my wall. How I lay
in wait for it, for what it told, and why
the window set it there! To steal a glance
I broke a penance of the eyes. Visible grace
flying in the air, the daily floodings in
and ebbings out of light were signs to be read
richly into, for I knew him by these things.

A residue of dry secrets. Those corrupt
picturings on the prie-dieu. Schoolmen and dust.
There is nothing to see, touch, taste, smell, hear.
I am a bell they left in disrepair
without a tongue. Looking back is my sin . . .
I have a narrow bed to meet despair,
pen and paper and blood to dare to tell
of a thing that I know. His indifference
is a stern mercy. The sleep of the just
is not different from the sleep of the lost.'

Coming Back

(for James Malpas)

When the something withdrawn (you cannot tell
exactly what or when) flows back into the blood
and you return from the damned into your own
(where living is at last to be living now
which the damned cannot know); when your beloved
is again beloved, and morning shows a tree
dressed in light, meaning and memory
at peace in its leaves; when your thought is cool
as linen and you go downstairs to receive
a letter from a friend with total recall
who tells you what you were, and you listen
with attention to rain on the skylight
which tells you what you are;
                        then you know
that nothing is so lost or gone to waste
that it cannot start again; as when you leave
the city for the pulling air and the sea
which turns upon itself and fills you, something
bows you down to the ground, bows you weeping
down to the sand, weeping there and giving thanks.

An Afternoon in the Parc Monceau

The weather was foreseen, but not the world
in its field, this was not dreamed up.
But there it was, one interwoven place
where the sun poured down and the children
were moving jewels. A poplar unfurled
its shimmering skein, the candles were dancing
in flocks on a chestnut; they danced in measure,
curtsied, kept balance. And between these things
was no division.

Half way through the afternoon, a wedding came
in slow procession through the crossing place;
solemn groom, troubled bride, for all of time
stiffly on the grass. But a smiling wind
unloosed them, and blew her long veil back,
and stretched it out and worried it and showed
her face light up, so animate and lovely
her hands nor his nor any hand could tame
the joyous flapping of that veil.

Lovers on the strong breathing ground,
intercede for us, kiss and kiss, obey
the irresistible thing, be tender
as the folded bird with his soft brown eye
who shares your bank of shade, or say
one motionless branch of becoming.
Beyond them in the half-lights, benched and blessed
between columns, the old sat on amazed
as one more summer climbed to its towers.

Rilke in Paris

For the shy young man from the provinces
life began seven floors up.
His days were a spiralling, down
the narrow stairs, past the acid concierge
on sentry duty at the door, and out
into a sunny vertiginous freedom.

That first autumn was a dream of autumn
unreeling along the intricate paths
of city gardens. Waiting for the winds
to bite into his art, he wrote long letters
to his quickly fading parents, figures
on a daguerreotype of the dead.

Plunged beneath green lamps in the library
his dream continued in the submarine sleep
of vacant faces and swaying heads;
desiring nothing more than this,
he took his place for life
on those benches of the blessed.

It happened on an ordinary day,
mid-week, when he was in a corner seat
of his favourite café. He was convalescing,
weightless, letting the steam invade his thought,
forgetting the mountains, the words on vellum,
his body anchored and his blood at work

from watching the shrunken lamplit street
when a woman he had seen before,
a silent neighbour and companion,
poured her wine to the point of overflowing
and went on pouring in the horrible growing
silence that rose like walls around her.

He knew he must call her, call her back
invisibly. Wearily he climbed his stairs,
through the sickening density of air
and like a swan reentering its element
he steadied his hand and began to write
with a cold eye under the icy lamp.

Entracte

Shadows of roses play beyond their grid,
and a slight deficiency in the blood,

a lassitude and nothing serious,
stales like wine where we loll on grass,

our Indian summer on an English lawn.
But a snowy man rises with distinction

and his royal progress to the door
takes a quarter of an hour

which is purpose. Where's the high-collar
starchy shirt-front chin-up character

that tied and retied its silk to perfection
and laid down its life? A whiff of cologne

in Central Africa. He never got there,
but taps the glass and checks the pressure,

writes the log-book, moves about the house
in socks and shoes, preserves a marble face

that cries for blood like a late Edwardian
waiting for something please God to begin.

In the Cinque Ports

At the edge of things, in a bungalow
by the sea, a barometer stuck at 'fair'
ignores the manic sky. It is unmoved
by changeable behaviour in the close
society of weathered ladies
at gin and bridge beneath its glass.
They are an angular, slender-legged
breed; and one by one they go missing.

Discreetly, the merciful visits begin
to their silken tents propped on pillows
and end fantastical in memory, in wraps
of white, like Great Aunt Vi, brought home
from crossing the Zambezi to a death bed
somewhere in Hythe. Her thin extended hand
was a beckoning claw, her piping voice
'come close, come closer child' a terror.

Their line is thinning out, a cluster
of feathered hats on the sea front
threatening to fly in a blast of air
that makes a tatter of tailored skirts.
A flapping row, strung out along the wall
that keeps the sea in place, with its light
blue horizon, unshakeably set fair,
beyond the churning salt and mire.

Wellingtonia

From his armchair in the home counties
grandpa followed our Asian journeys
with an atlas of the world and supplement
of postcards, hasty and irregular,
from Delhi, Cochin, the southern tip of India.
He plotted the route with eccentric care
as if he'd travelled it. But when I came back
taller, with the deferent condescension
of the barely adult, he missed the boy
back in the kitchen playing Snap with Ada.

Ada ruled the larder. My life was pale yellow.
There was bread and butter, Gentleman's Relish,
a drawer of knives, a row of yellow bells,
two sisters and a blue Wellingtonia.
We gathered up dead needles by the handful,
but the tree renews them, said grandpa,
full of facts and detail. In the war
he drilled his soldiers on the tennis lawn
or so my mother said. I liked to linger
in the shadow of a pine green corridor.

I felt the velvet drapes and smelt the turpentine
but loathed the girlish shoes I had to pose in.
Granny was an upstairs painter, her studio
an attic. Downstairs was dark, the conversation
social: families, marriages, property, sport.
Once under drugs, granny rose from her ground-floor
death bed and with uncontrollable strength
began to mount the stairs. She was found
in the small hours, more than half-way up,
scrabbling at a window as if for air.

She spurned her nurses, dreading helplessness.
Prodigal of flowers, her lacquered garlands
live with us on tables, beds and chairs.
A morning fire below the smoky mirror,
a mixture of lights, tonic in a glass

can reassemble her to me. The Wellingtonia
dropped its skirts of branch almost to the lawn,
extinguishing the grown-up room in shadow.
The turn in the corridor was always dark.
Grandpa packed her portrait and moved on.

Later, he found he shared an interest
with his otherwise disappointing grandson
—bookish, secretive, no good at sports—
and that was pipe-smoking. We would light up
together, a rosewood briar and a meerschaum,
breathe in with the little pop of fishes,
tamp the tobacco down and move off slowly
crunching round the gravel drive. I still
have the last of the Three Nuns and cleaners.
My other achievement was a motorcycle.

Nénuphars, Nymphéas

(for David Gascoyne)

This the swollen bandaged hands have done
this delivering up to waters
to the lily's yellow air with absolution.

Absence in a cup, glowing on the waters,
signals from a bed of pale island green
to absence's reflection where the hard head falters;

except this heart's white face was seen
shy mouth pressed by the window's cold,
the question what could lilies mean

so drenched in glass, or how unfold
one surface leaf, is diverlike his work,
his downward knowledge of gloom and gold.

The Well

Sick of talk, we cycled down the green tunnel
of altering lights to that village where the well

is just an opening in the middle of the road.
There was no one about, and it was good

to splash our foul words off. It's the chain
of what we do not mean, up against the skin,

scoring cuts where there were none . . . In that village
the water stood so clear I could not gauge

its level until I dipped a finger in.
I tried, and failed, to see how we remain

transparent like that, as if our modest god
who has no tongue could tell. I know there is a flood

that moves with us. That it is in the flood, at swim
from you to me. That it may not have a name.

Magpies

There's speech for this only when she's away
and the big-tailed premonitory magpie
keeps his own bad company

creaking through the thick
sky, and no easy talk
will do, and no high talk

of marriage keeping guard
over another's solitude
—this is solitude

and horribly unlike the medium
we sustain through time;
I cannot name

it, lived in not seen
or known
only as a flickering third companion

is known, or by its rupture,
a whirlpool in the running water
that was clear

and went grey with the breath
of being looked at, the ghost of faith
that went in the space of a breath

—and came back, found its way in
like the rain
—but the window wasn't open—

like her then, opening the door
to join this bedraggled bird, changed by her
into two, and possibly three, or four.

Colours for Thomas

We lay together under the gold
curtained window where cobalt travelled
in squares up the wall, exact, controlled

by the scheme that closes orange heat
inside a fringe of violet.
Those squares of blue were desolate,

they sucked us dry;
all that afternoon we lay
stranded in our expectancy.

Nothing can alleviate
a sky so taut,
but sit it out

—it will break for sure
into its reconciling colour.
The day is softened at either

end, so we walked at sundown
when the pink clouds brood on the green
leaves. You showed me again

how crimson lives in the emerald grass,
shadow to its central brilliance,
heart of its ecstasy, the changeless

deathlike tone. The end of light
is constant, mixed on every palette
even as we mix against it

with passion and accident and reason.
Under a window lay our newborn son
crowned by a spectrum, the seven strands of vision.